はしがき

天災・人災、被害者の身元確認情報に効果を発揮するのが、この靴であり、この靴の構成・用途・使用方法などを独創的に表現する。例えば、歯型などで身元を確認する方法は周知の通りである。しかし、靴で身元を確認する方法の表現は例が見られないので、これをイラスト解説で著作権の表現をする。

上記が初版であったが、足首に紐類を設ける構成の表現は本来の表現であり、初版に掲載できなかったワカール本来の表現を解説する。

Preface

These shoes demonstrate an effect to a natural disaster and a man-made disaster, and a victim's body identification information, and it expresses creatively the composition, the use, the directions, etc. for these shoes.

For example, the method of checking an identity with a denture mold etc. is well known.

However, since an example is not seen, expression of a method which checks an identity with shoes expresses copyright for this by illustration description.7

The above was the first edition, but an expression of the composition which sets up string kinds to an ankle explains an expression of the WAKARL origin which couldn't be an original expression and carry it in the first edition by additional publication.

目　次

１、天災・人災 被害者の身元確認情報 情報携帯 ワカール(イラスト解説)

⑴　紳士靴---------- 9

⑵　女性靴----------10

⑶　運動靴----------11

⑷　長靴----------12

⑸　作業靴----------13

⑹　登山靴----------14

⑺　デッキシューズ----------15

⑻　スキー靴----------16

２、身元確認情報　ワカール　Ｗａｋａｒｌの仕様

⑴　表----------17

ゴムリング、黒、天然ゴム・ナイロン、耐水紙、二つ折

⑵　裏

中央に粘着剥離紙

⑶　表面の構成----------18

ゴムリングは表裏面移動自在

折り込まれた面を開く----------19

⑷　身元確認情報（個人情報）記入用紙

日本語と英語の記入欄(11.6×3.5)

３、ワカールの使用

⑴　紐付きの靴----------21

ベロの内側に剥離紙を剥がして、粘着面で貼り付ける

(3)--29

Sports shoes

(4)--30

Boots

(5)--31

Work shoes

(6)--32

Mountain-climbing boots

(7)--33

Deck shoes

(8)--34

Ski boots

6、Identification information The specification of WAKARL

(1)--35-

Front

Rubber ring, black, natural rubber nylon, waterproof paper and folding into

two

(2)

Reverse side

Central adhesive tape

(3)--36

Surface configuration

Both sides move and are a rubber ring freely.

An inserted form is opened.

(4)--37

Identification information (personal information) entry form

Frame of Japanese and English(11.6×3.5)

7、Use of WAKARL

(1)--39

Strings-attached shoes

Peel peeling on the inside of the tongue and paste it on the sticky side.

(2)

Shoes without string

Paste on the inside of the tongue between Belo and the rubber band

It's stuck to the inner side in case of high-heels.

(3)--40

Attachment of WAKARL

A link is set in an ankle.

8、Tow of WAKARU

(1)--41

A tension gage is hung on ankle string of WAKARU.

(2)

1.5 kilograms of traction

(3)--42

1.6 kilograms of traction

(4)--------------------42

1.8 kilograms of traction

(5)--------------------43/44

2.0 kilograms of traction

Adhesion of a tape can be taken from more than 2.0 kilograms of traction.

The state that WAKARL came off from shoes

１、天災・人災 被害者の身元確認情報靴 情報携帯靴 ワカール（イラスト解説）

靴の中に個人情報カードをセットで身元確認

⑴ 紳士靴

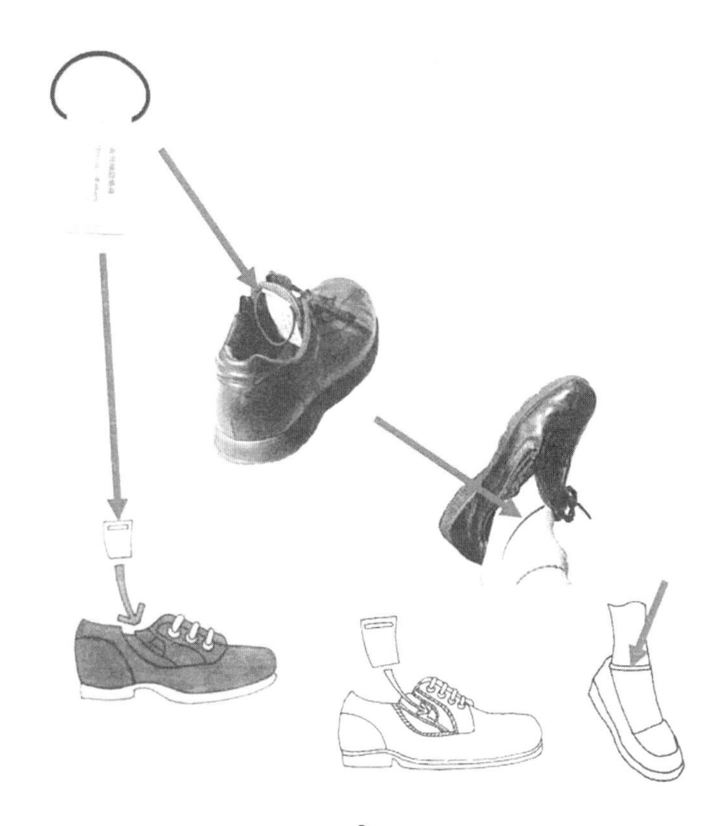

⑵　女性靴

⑶　運動靴

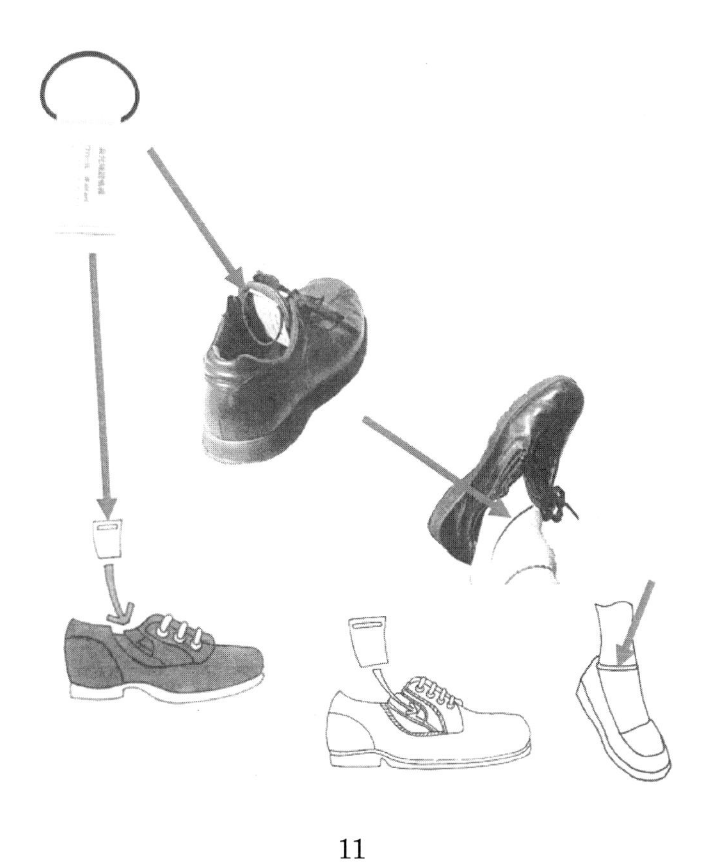

⑷　長靴

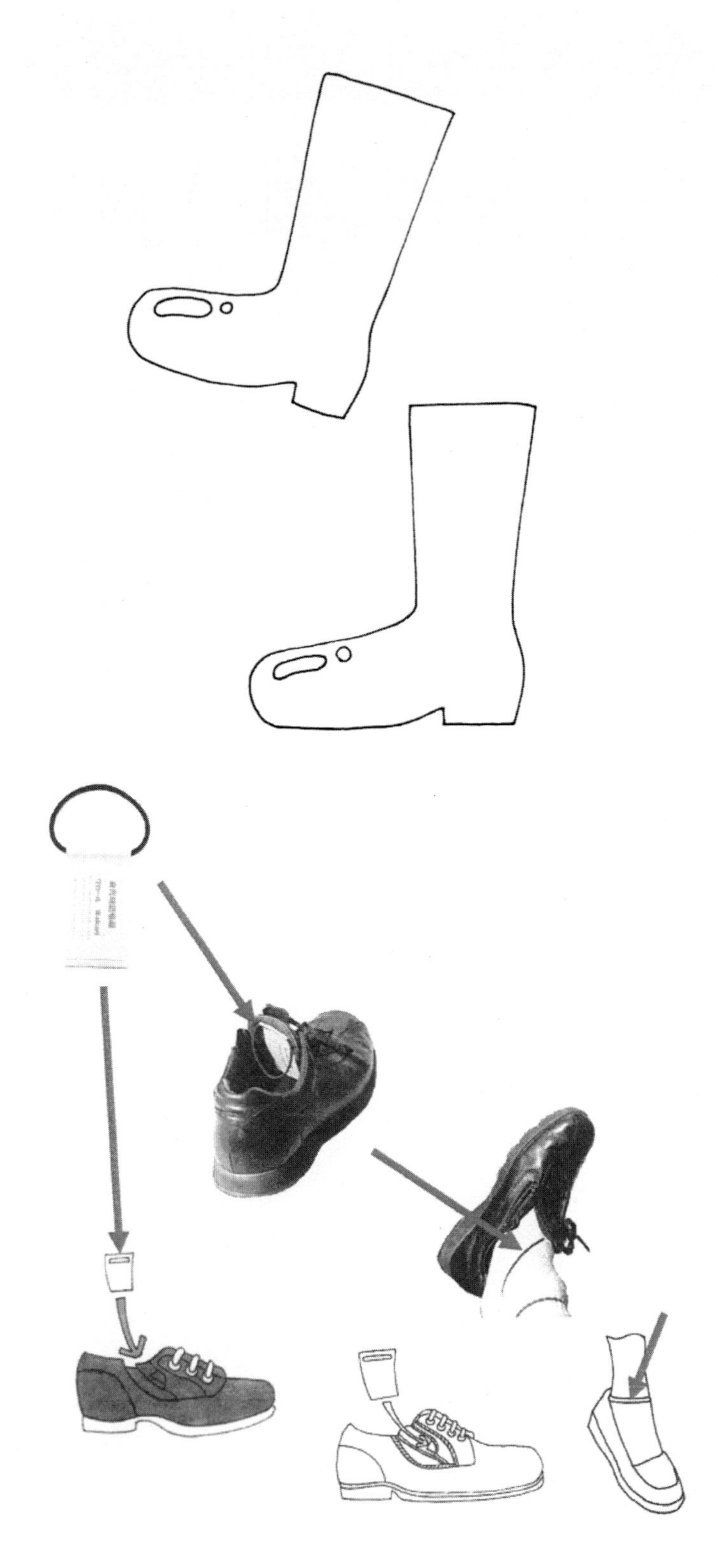

⑸ 作業靴

生き埋め・土砂崩れ

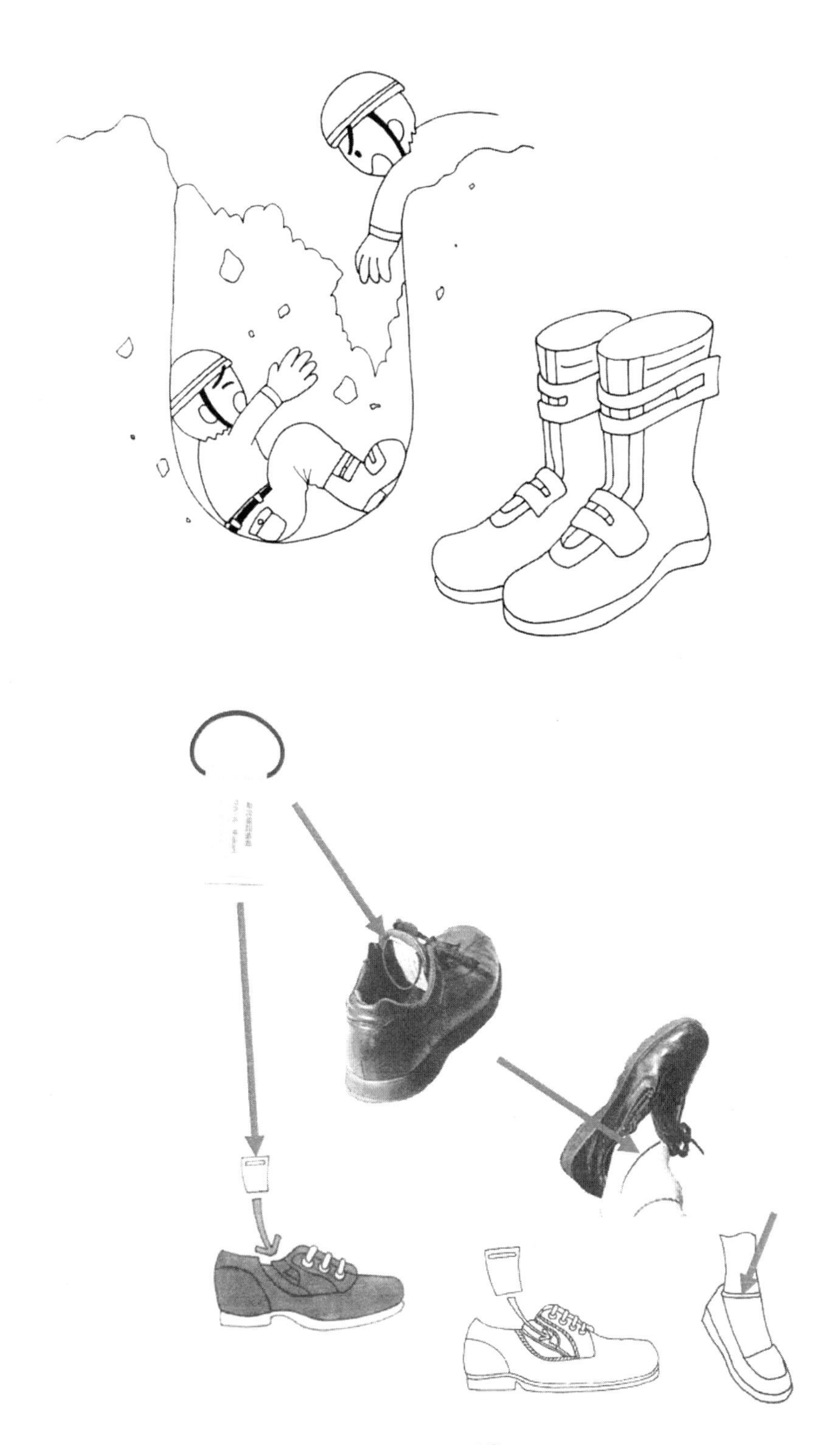

⑹　登山靴

遭難

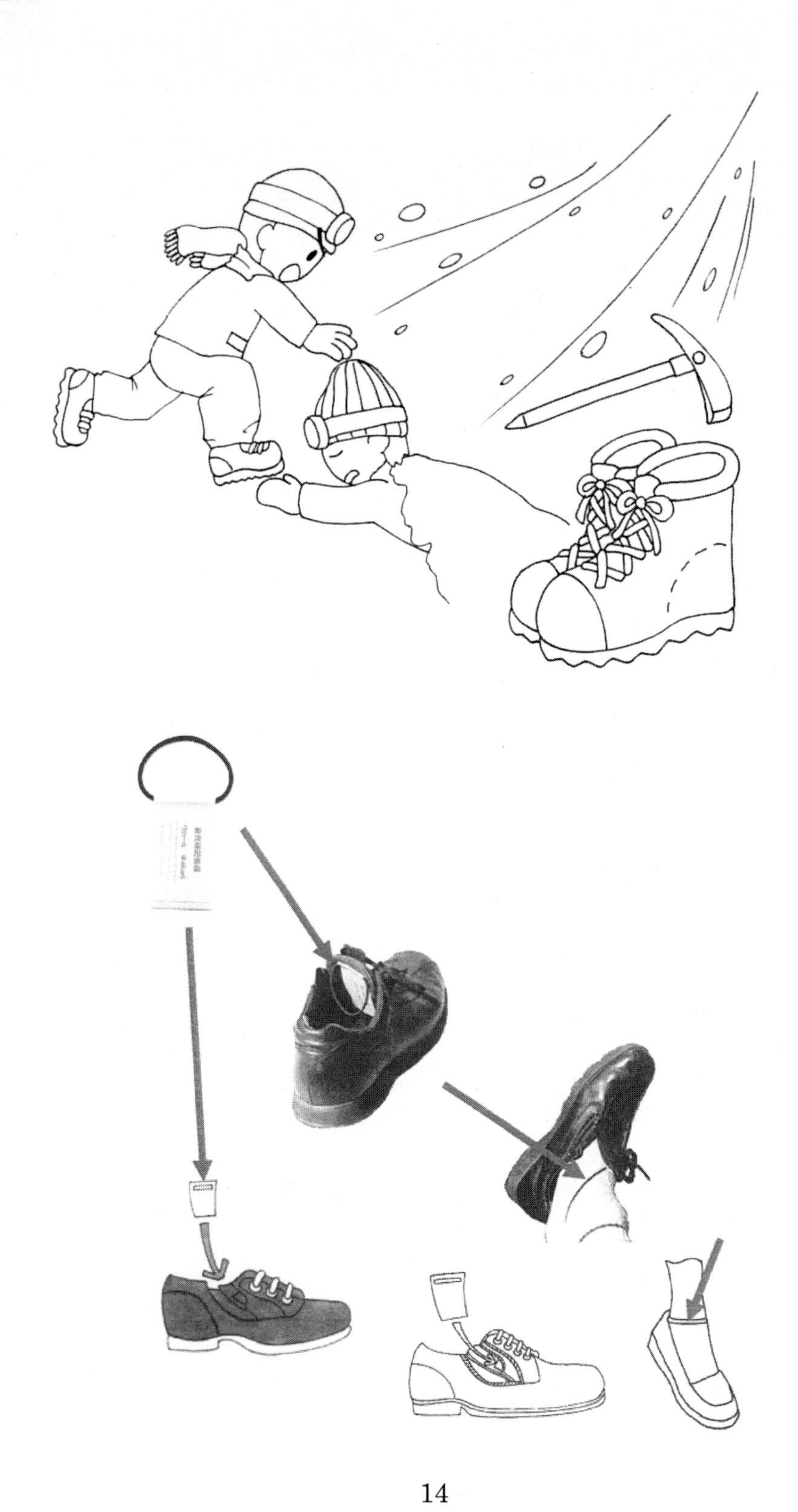

⑺　デッキシューズ

転覆

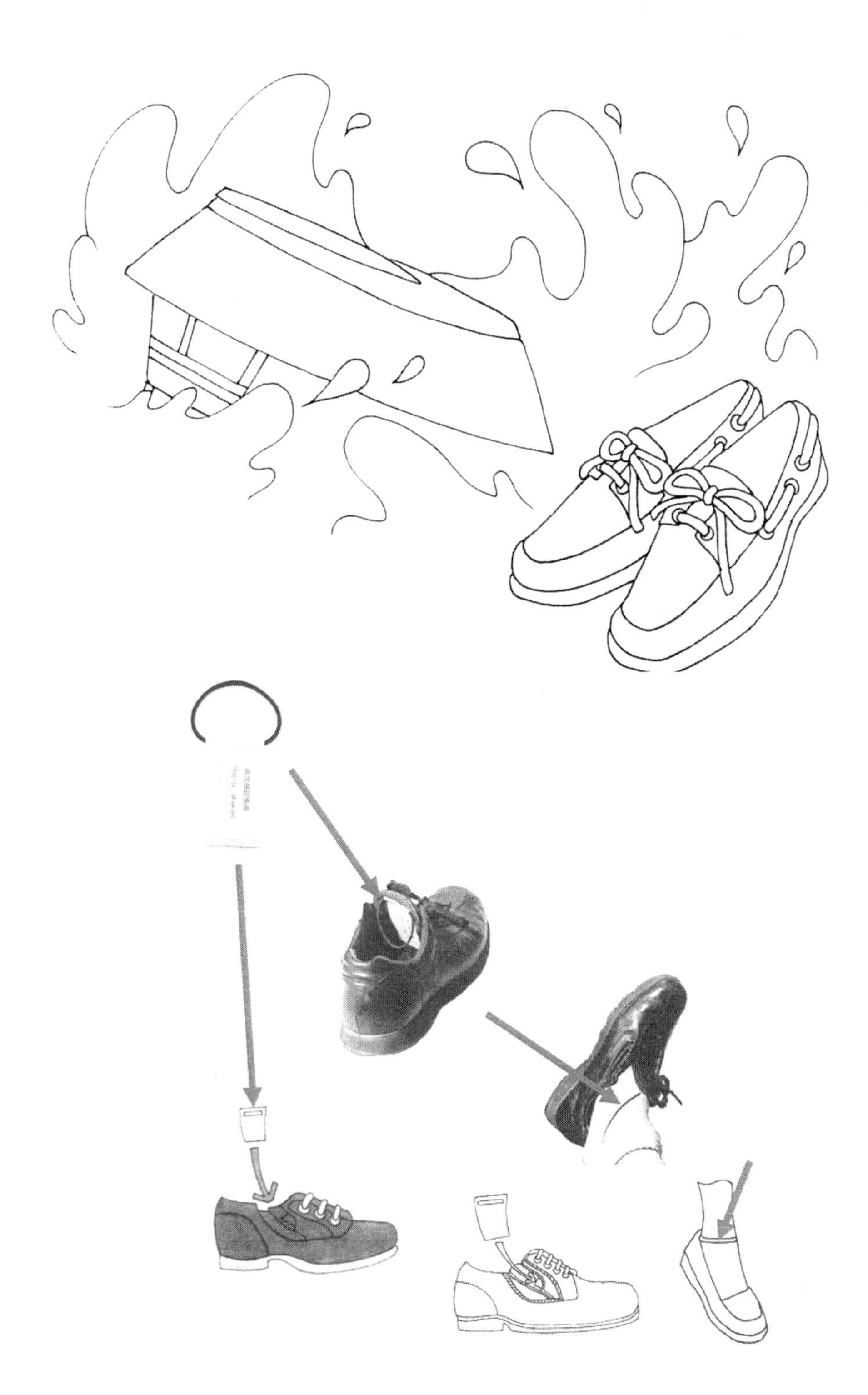

⑻　スキー靴

雪崩

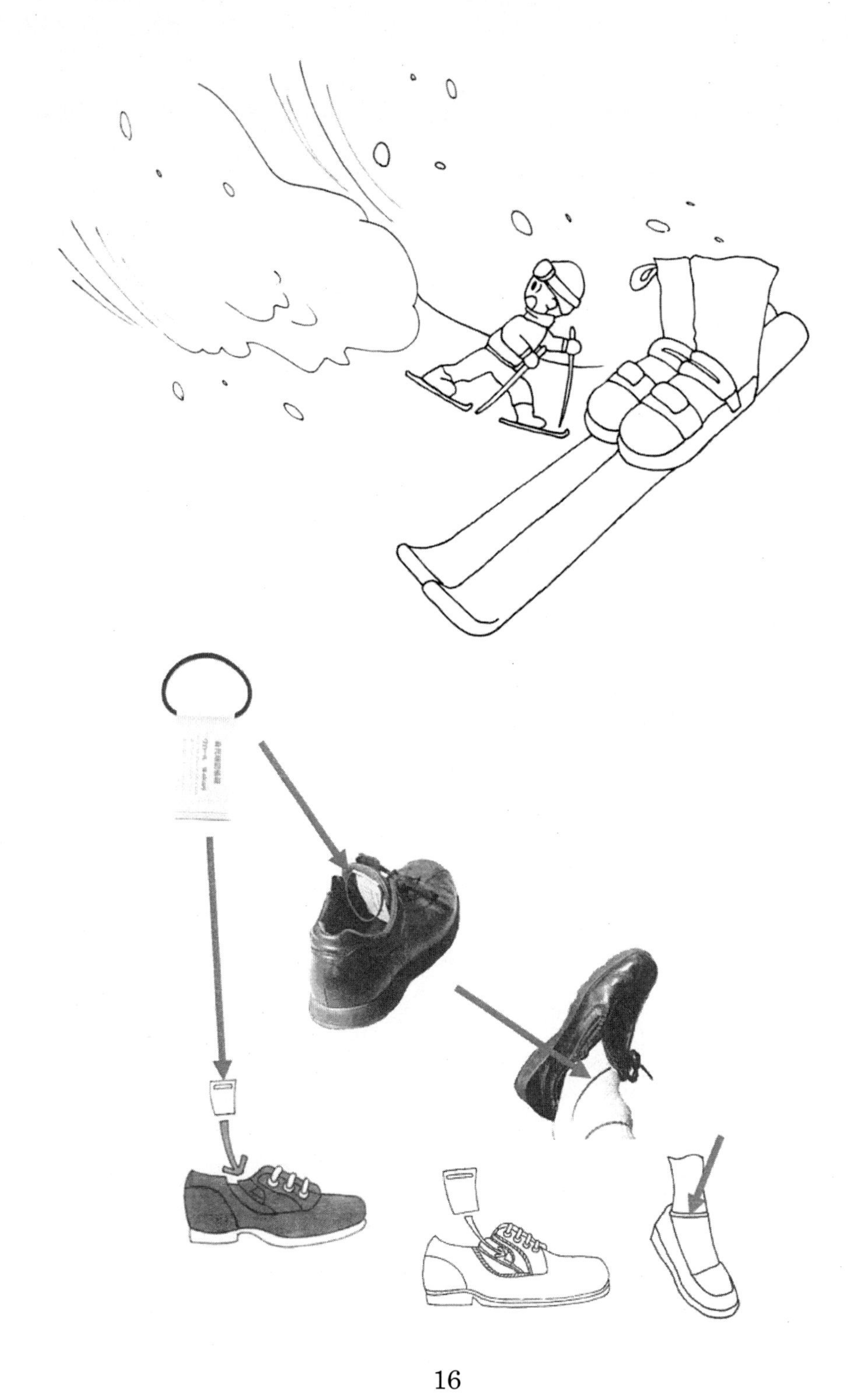

２、身元確認情報　ワカール　Ｗａｋａｒｌの仕様

⑴　表

ゴムリング、黒、天然ゴム・ナイロン、耐水紙、二つ折

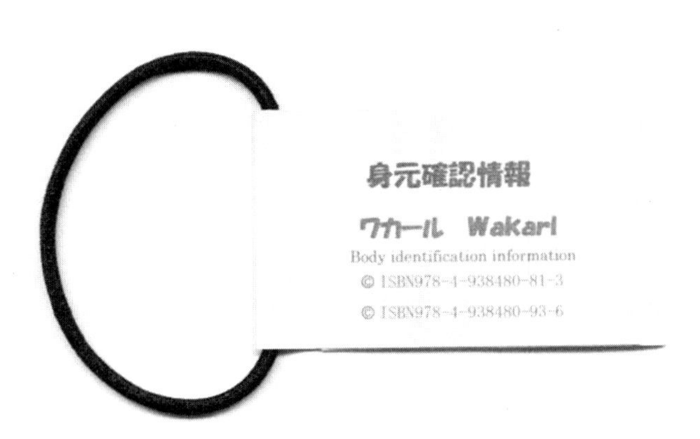

⑵　裏

中央に粘着剥離紙

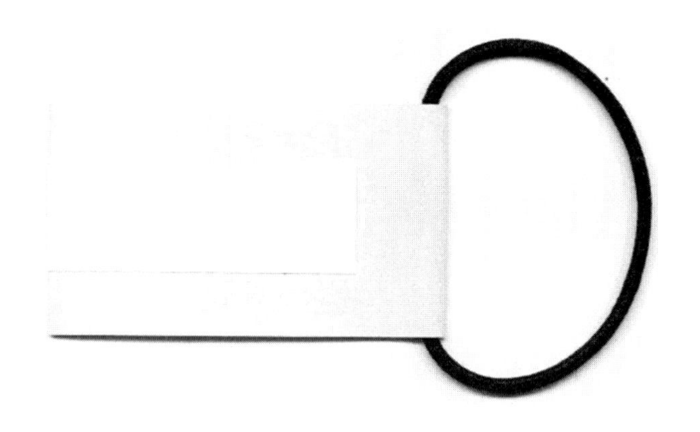

⑶　表面の構成

ゴムリングは表裏面移動自在

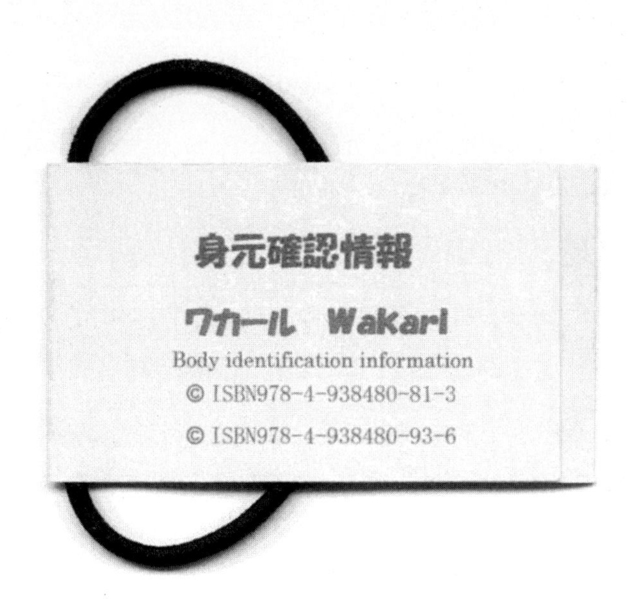

折り込まれた面を開く

身元確認情報

ワカール　Wakari

Body identification information

© ISBN978-4-938480-81-3

© ISBN978-4-938480-93-6

男　女/MAN LADY

SEASE

⑷　身元確認情報（個人情報）記入用紙

日本語と英語の記入欄(11.6×3.5)

名　前 NAME	男　女/MAN LADY
住　所 ADDRESS	
電　話 TEL	
病　名 THE NAME OF A DISEASE	
血液型 BLOOD	
その他 OTHERS(個人番号)	

3、ワカールの使用

⑴ 紐付きの靴

ベロの内側に剥離紙を剥がして、粘着面で貼り付ける

⑵ 紐無しの靴

ベロとゴム帯の間に於いて、ベロの内側に剥離紙を剥がして、粘着面で貼り付ける

ハイヒールの場合は、内側の側面に貼り付ける

⑶　ワカールの装着

リンクを足首にセット

4、ワカールの牽引

⑴ ワカールの足首紐にテンションゲージを引っかける

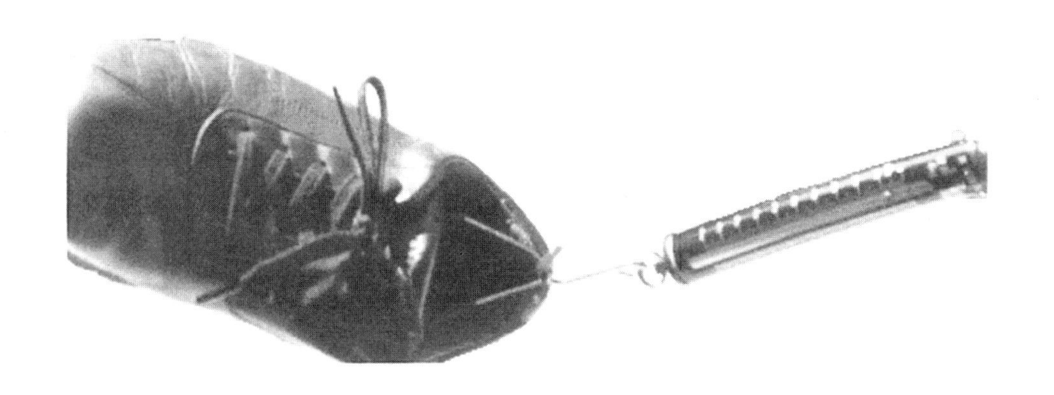

⑵ 牽引力1.5キロ

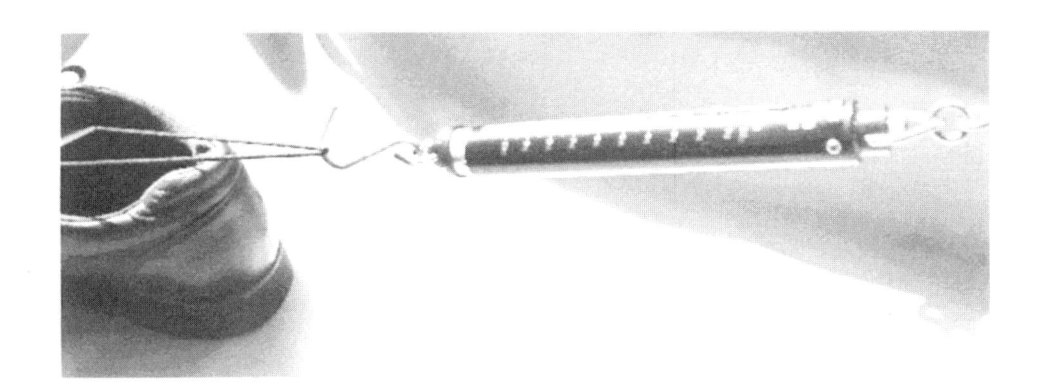

⑶　牽引力 1.6 キロ

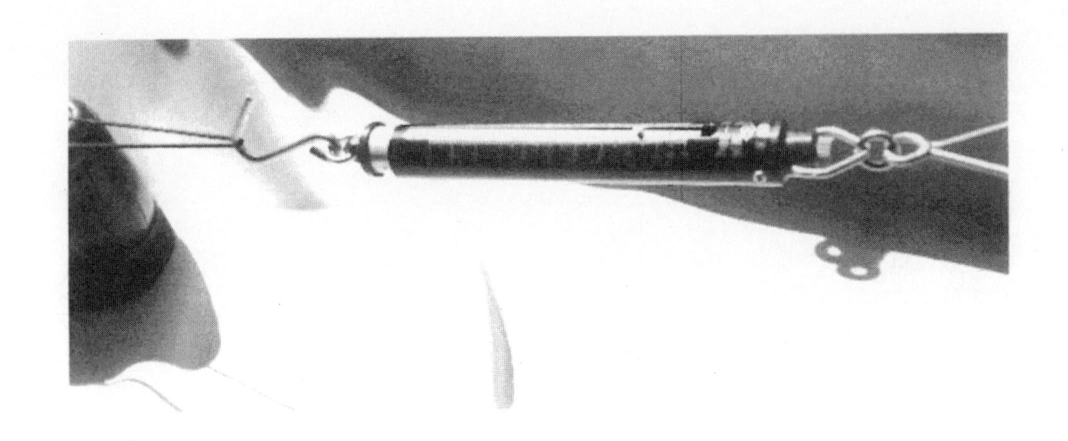

⑷　牽引力 1.8 キロ

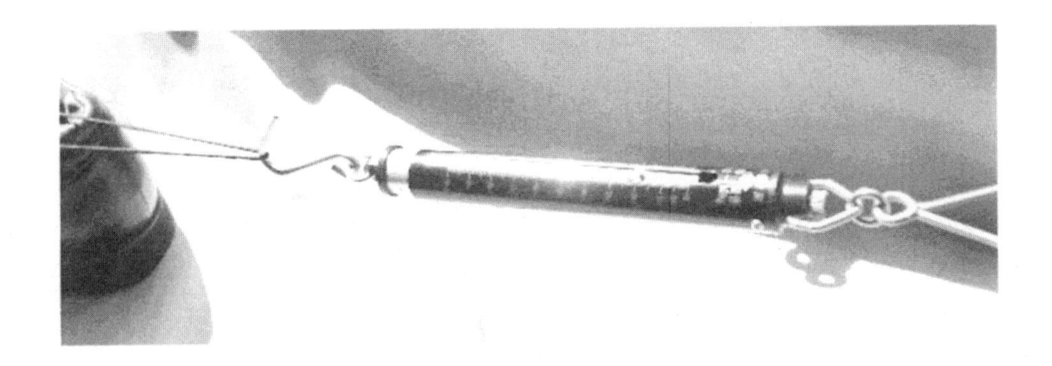

⑸　牽引力 2.0 キロ

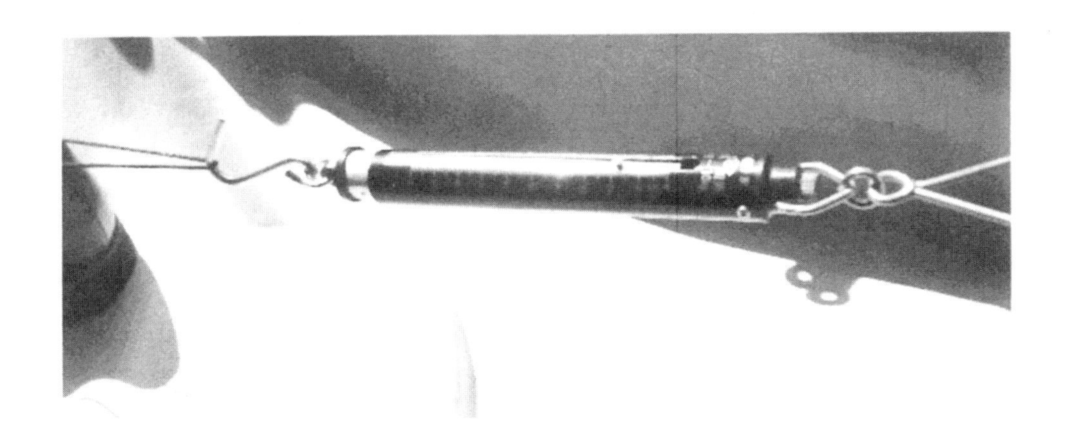

牽引力 2.0 キロ以上からテープの粘着が剥がれる。

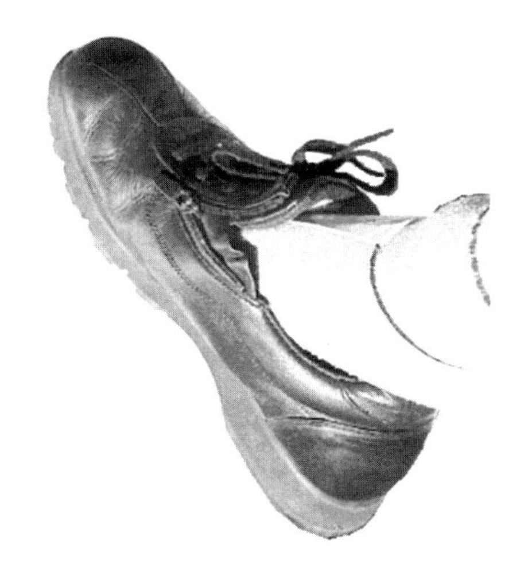

牽引力 2.0 キロ以上からテープの粘着が剥がれる。

ワカールが靴から剥がれた状態

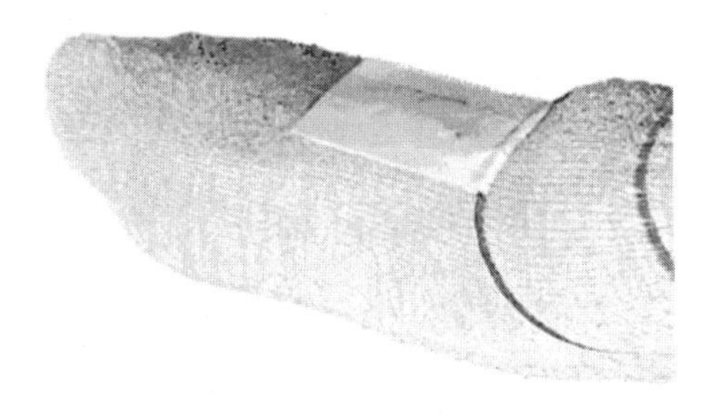

2、English of the usage

A natural disaster and man-made disaster　Victim body identification information shoes Portable information shoes Ｗａｋａｒｌ

(Illustration explanation)

The body of a personal information card is identified by the set in shoes.

(1) Gentleman shoe

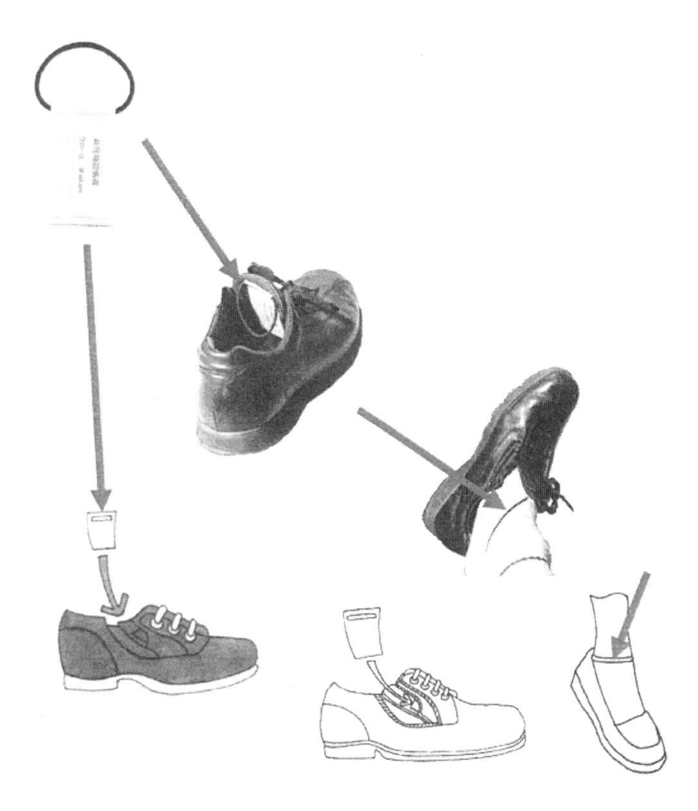

(2) Female shoes

(3) Sports shoes

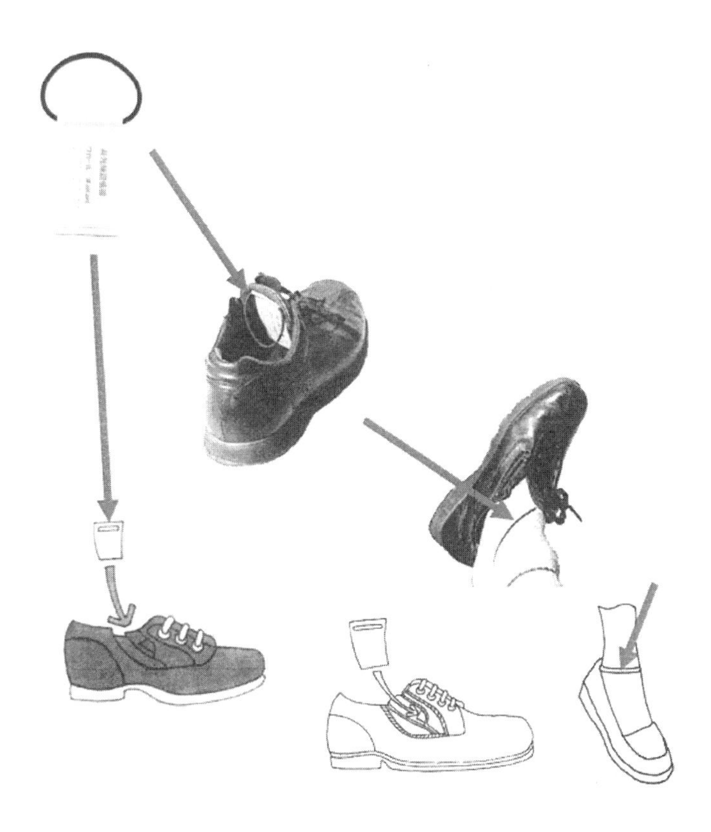

(4) Boots

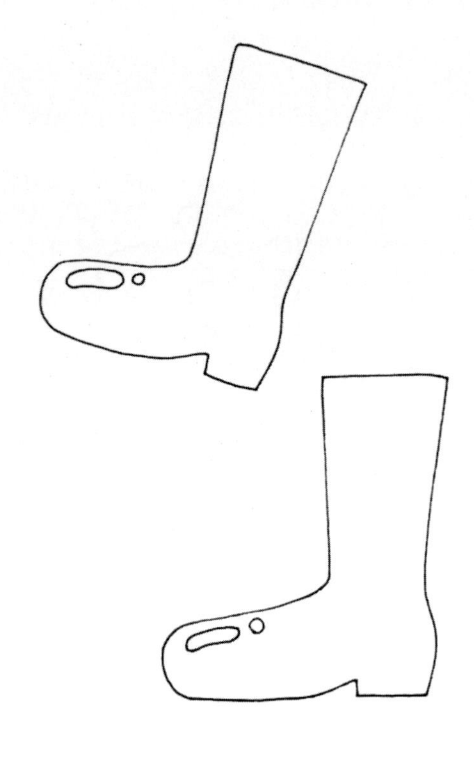

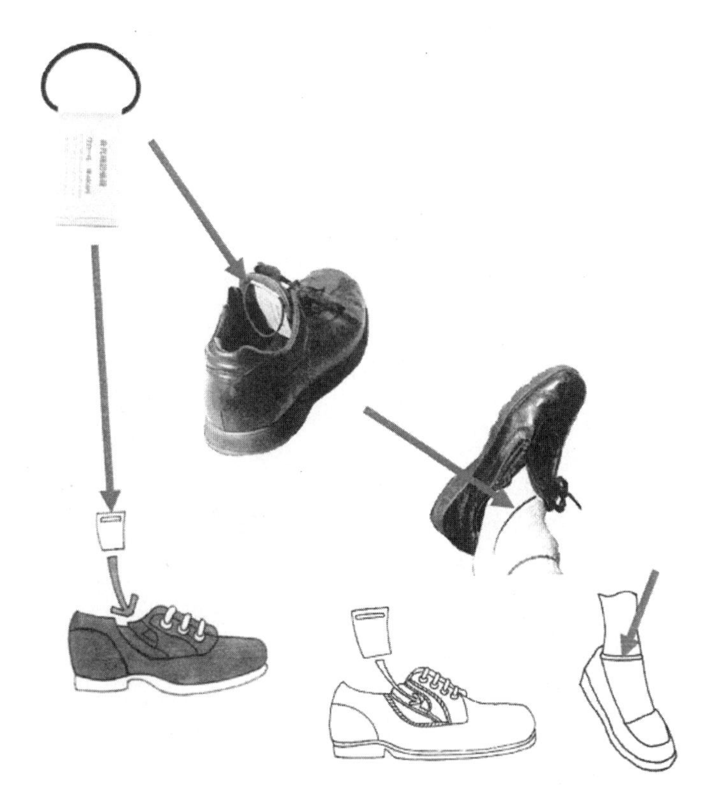

(5) Work shoes

Mudslide

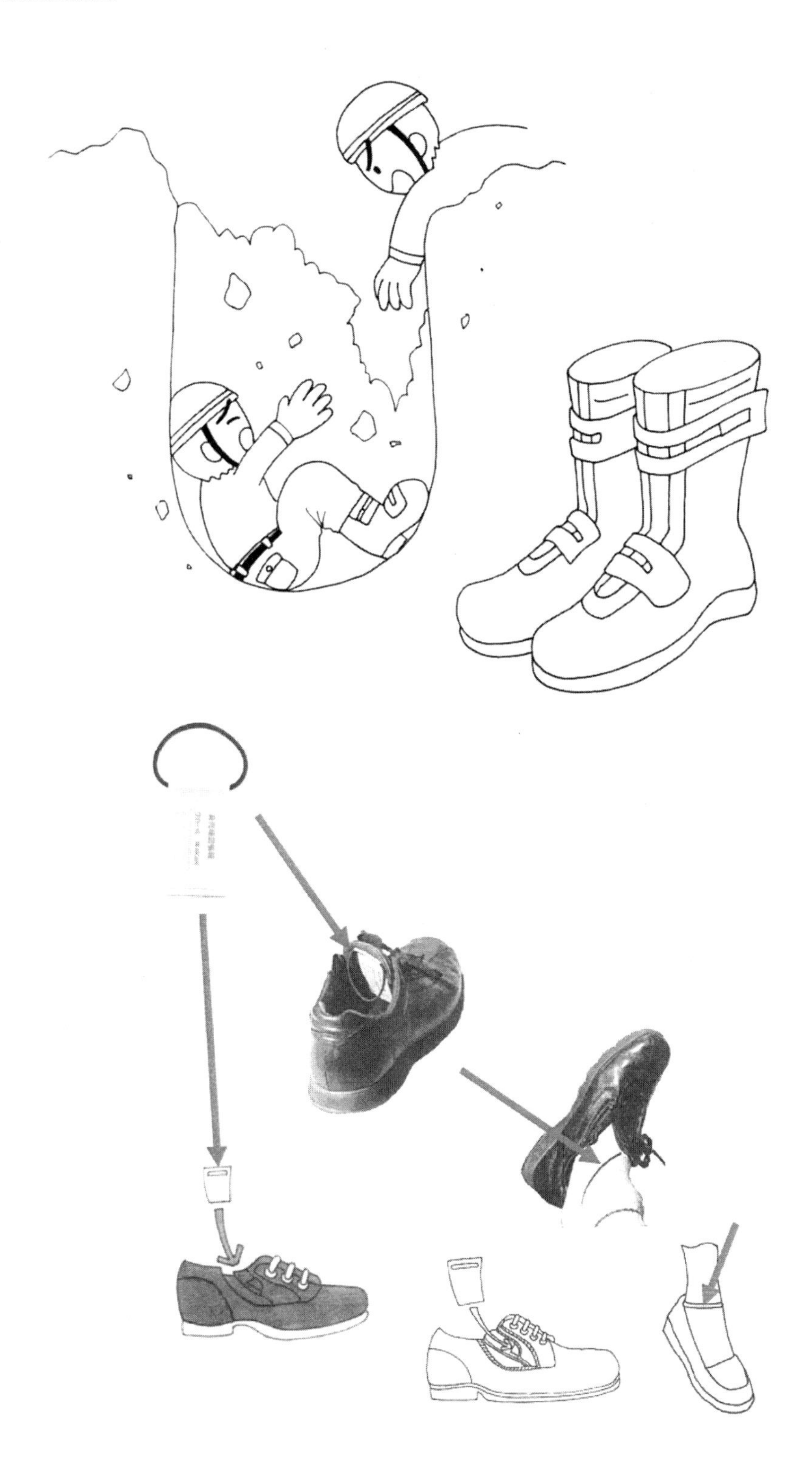

(6) Mountain-climbing boots

disaster

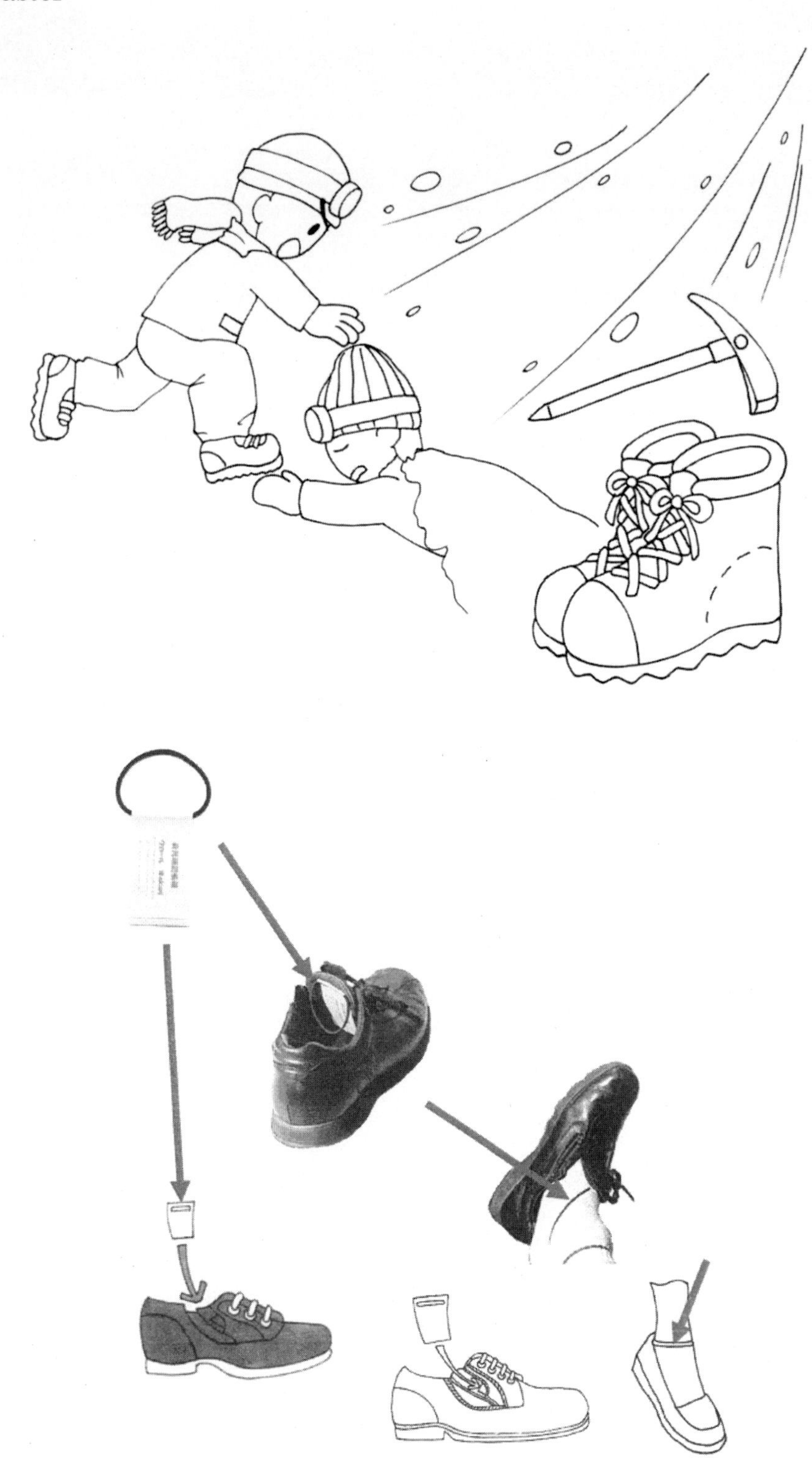

(7) Deck shoes

Overthrow

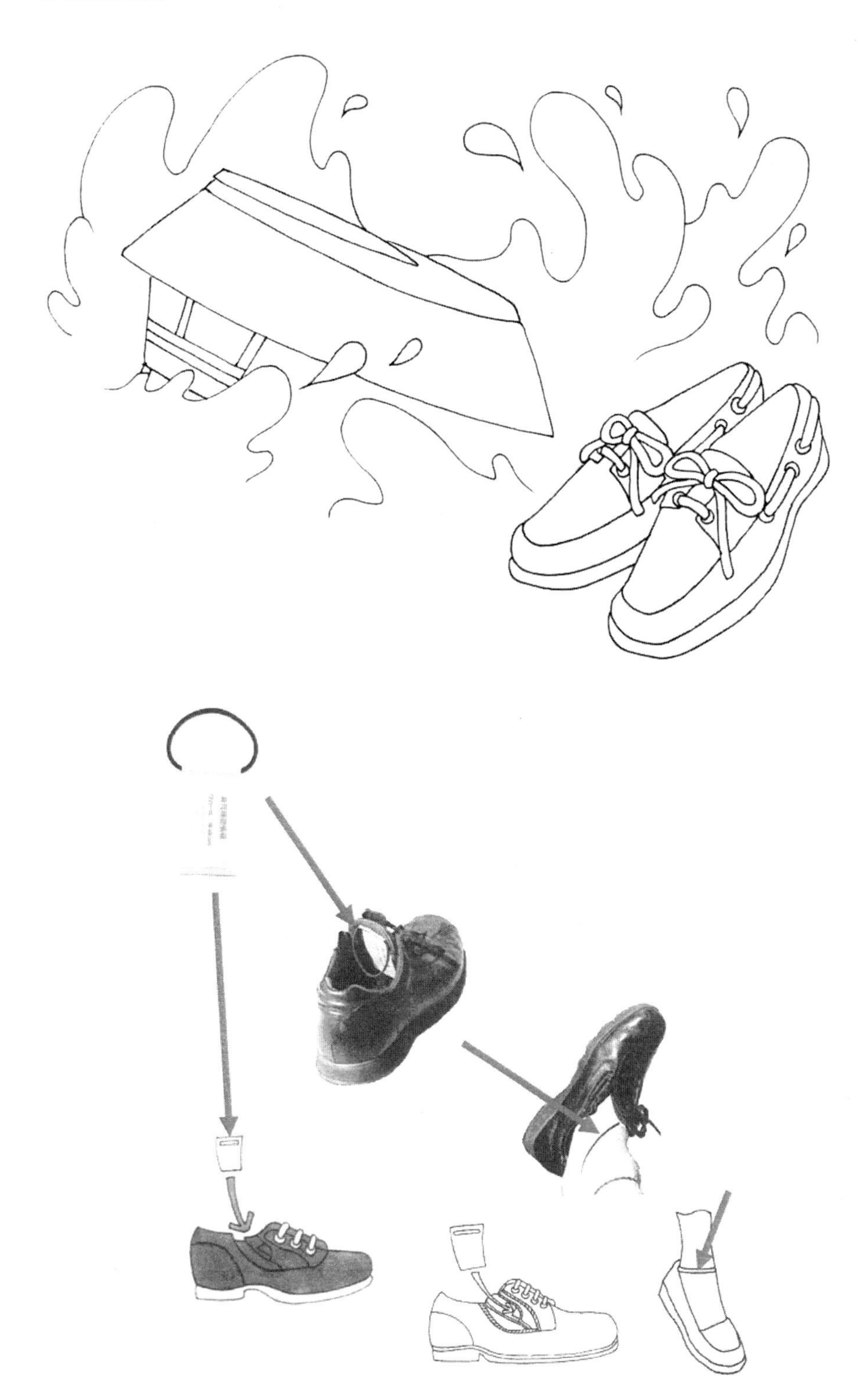

(8) Ski boots

Snowslide

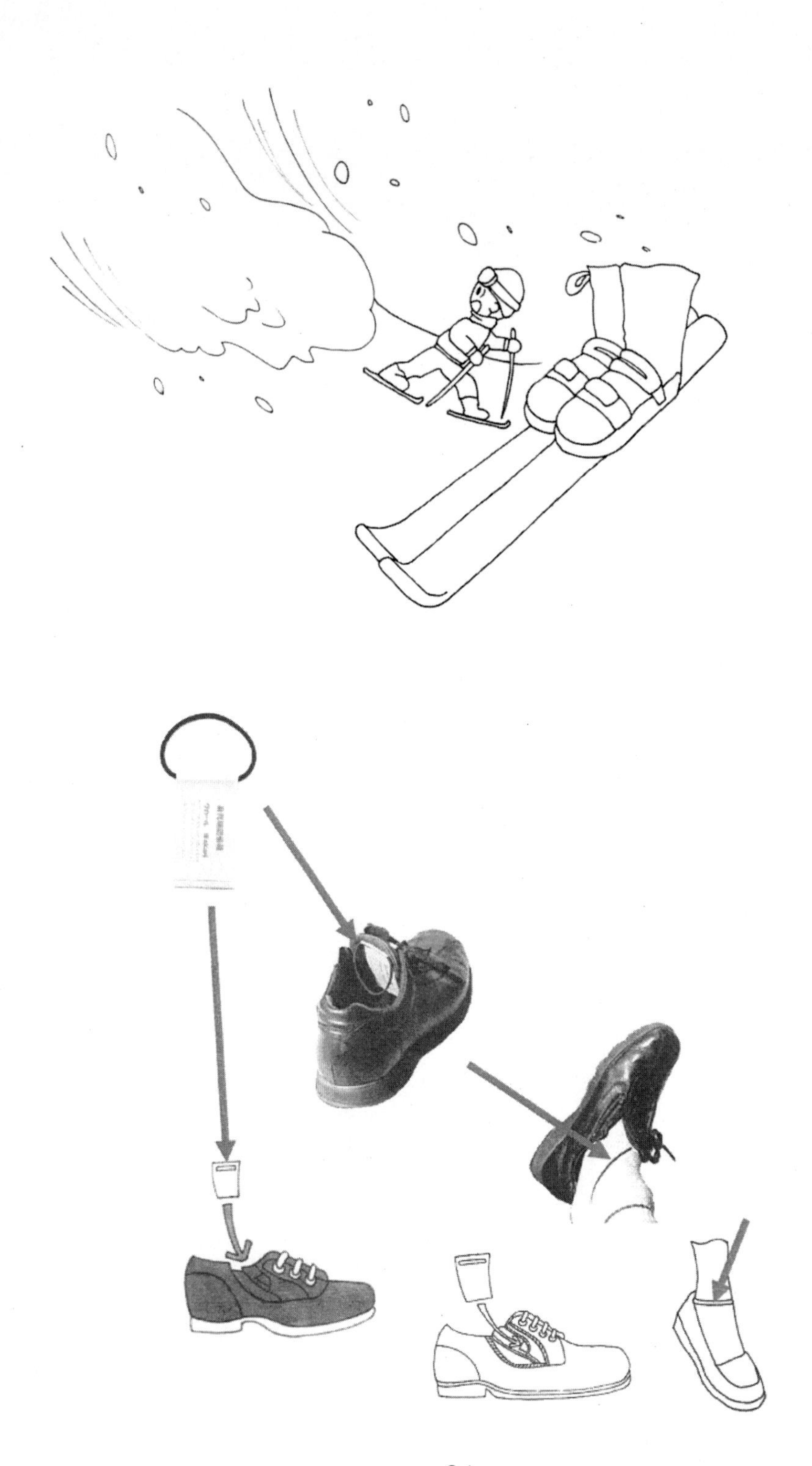

６、Identification information　The specification of WAKARL

⑴

Front

Rubber ring, black, natural rubber nylon, waterproof paper and folding into two

⑵

Reverse side

Central adhesive tape

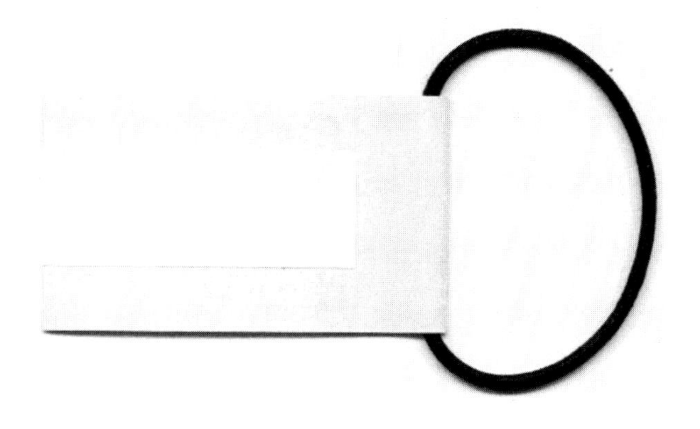

(3)

Surface configuration

Both sides move and are a rubber ring freely.

An inserted form is opened.

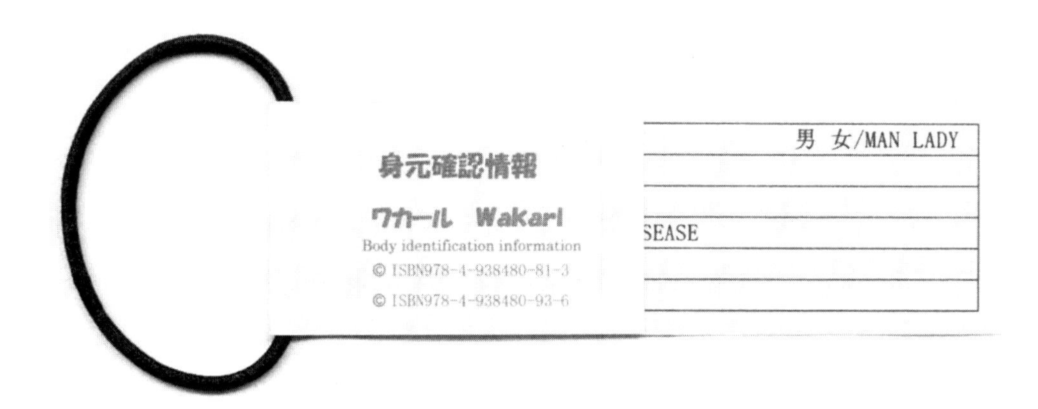

(4)

Identification information (personal information) entry form

Frame of Japanese and English(11.6×3.5)

名　前 NAME	男 女/MAN LADY
住　所 ADDRESS	
電　話 TEL	
病　名 THE NAME OF A DISEASE	
血液型 BLOOD	
その他 OTHERS(個人番号)	

7、Use of WAKARL

(1)

Strings-attached shoes

Peel peeling on the inside of the tongue and paste it on the sticky side.

(2)

Shoes without string

Paste on the inside of the tongue between Belo and the rubber band

It's stuck to the inner side in case of high-heels.

(3)

Attachment of WAKARL

A link is set in an ankle.

8、Tow of WAKARL

(1)

A tension gage is hung on ankle string of WAKARL

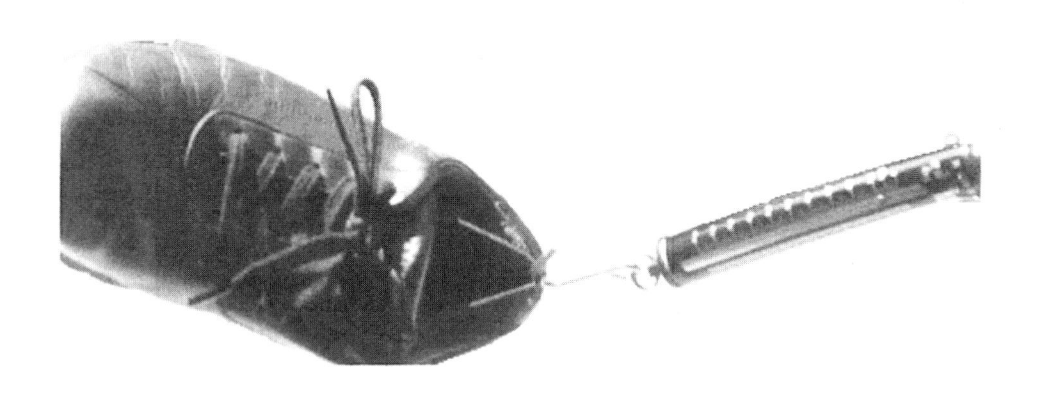

(2)

1.5 kilograms of traction

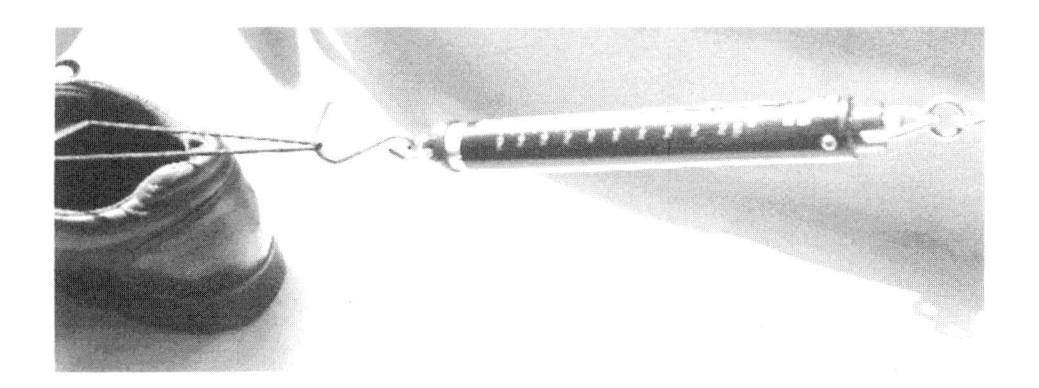

(3)

1.6 kilograms of traction

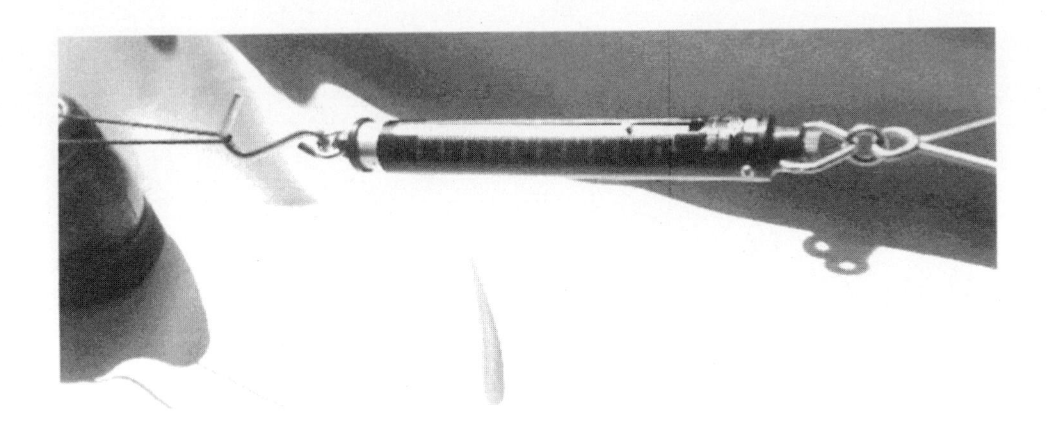

(4)

1.8 kilograms of traction

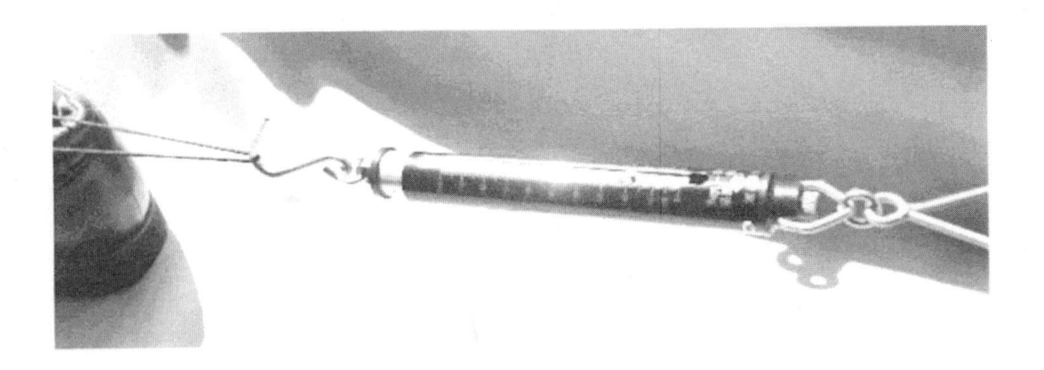

(5)

2.0 kilograms of traction

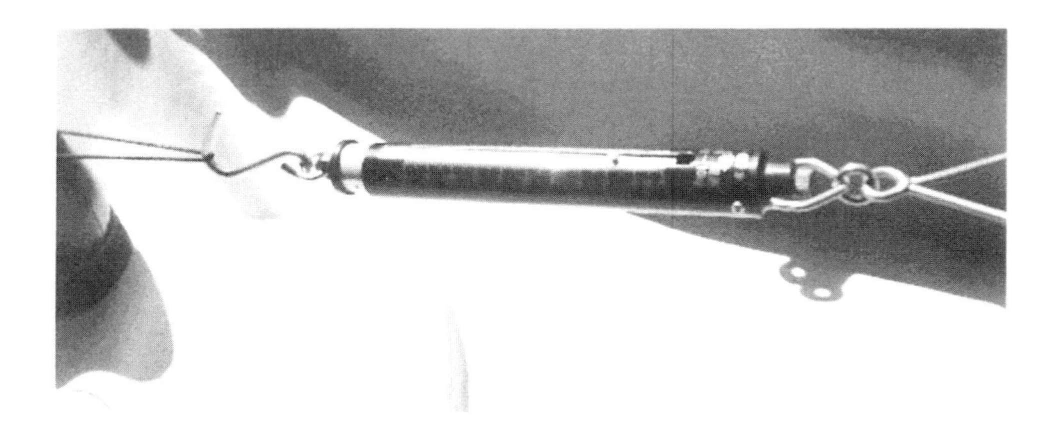

Adhesion of a tape can be taken from more than 2.0 kilograms of traction.

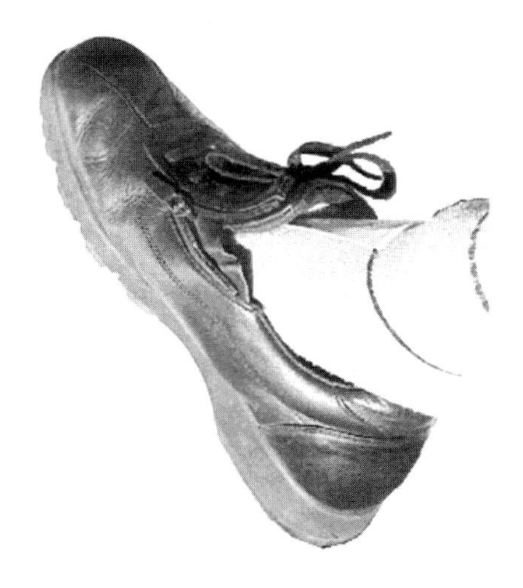

Adhesion of a tape can be taken from more than 2.0 kilograms of traction.

The state that WAKARL came off from shoes

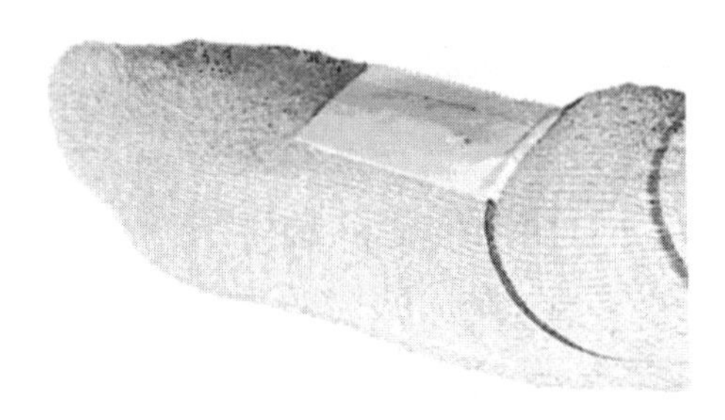

あとがき

「ナゼ？」

旅から帰る自宅周辺に来ると、気持ちが“ホッ”とする気持ちはナゼ？

旅行から自宅に帰る安心する気持ち、それは逆に考えると、事故、もしくは事件にあう事が気持ちの中に入っていると思う。だから自宅に帰ると安心する気持ちが沸いてくる。

現在、私達が皆が旅行、または会社に行く途中、何らかの事故・事件・病気などに身元が早く解かるのは靴。

それによって、家族・親戚に連絡が出来る特許の靴が誰も、自然な形で利用される事が大きな災害の時に大きな役割を果たす。

著者　斉藤　通
（さいとう とおる）

天災・人災 被害者の身元確認情報　身元・情報携帯 ワカール

定価（本体 1,500 円＋税）

２０１５年（平成２７年）３月３１日発行

No. 
ＳT93-6-017

発行所　IDF(INVENTION DEVLOPMENT FEDERATION)
発明開発連合会®
ﾒｰﾙ 03-3498@idf-0751.com　www.idf-0751.com
電話 03-3498-0751㈹
150-8691 渋谷郵便局私書箱第２５８号
発行人　ましば寿一
著作権企画　IDF 発明開発(連)
Printed in Japan
著者　斉藤　通 ©

初版、２０１３年（平成２５年）９月１２日発行に記載できなかった原稿の追加発行